Making Environmental Policy

T0273122

Making Environmental Policy

Two Views

Irwin M. Stelzer
and Paul R. Portney

The AEI Press

Publisher for the American Enterprise Institute
WASHINGTON, D.C.
1998

Distributed to the Trade by National Book Network, 15200 NBN Way, Blue Ridge Summit, PA 17214. To order call toll free 1-800-462-6420 or 1-717-794-3800. For all other inquiries please contact the AEI Press, 1150 Seventeenth Street, N.W., Washington, D.C. 20036 or call 1-800-862-5801.

ISBN 978-0-8447-7116-8
ISBN 0-8447-7116-3

1 3 5 7 9 10 8 6 4 2

© 1998 by the American Enterprise Institute for Public Policy Research, Washington, D.C. All rights reserved. No part of this publication may be used or reproduced in any manner whatsoever without permission in writing from the American Enterprise Institute except in the case of brief quotations embodied in news articles, critical articles, or reviews. The views expressed in the publications of the American Enterprise Institute are those of the authors and do not necessarily reflect the views of the staff, advisory panels, officers, or trustees of AEI.

THE AEI PRESS
Publisher for the American Enterprise Institute
1150 17th Street, N.W., Washington, D.C. 20036

Contents

Foreword

This volume is one in a series commissioned by the American Enterprise Institute to contribute to the debates over global environmental policy issues. Until very recently, American environmental policy was directed toward problems that were seen to be of a purely, or at least largely, domestic nature. Decisions concerning emissions standards for automobiles and power plants, for example, were set with reference to their effect on the quality of air Americans breathe.

That is no longer the case. Policy makers increasingly find that debates over environmental standards have become globalized, to borrow a word that has come into fashion in several contexts. Global warming is the most prominent of those issues: Americans now confront claims that the types of cars they choose to drive, the amount and mix of energy they consume in their homes and factories, and the organization of their basic industries all have a direct effect on the lives of citizens of other countries—and, in some formulations, may affect the future of the planet itself.

Other issues range from the management of forests, fisheries, and water resources to the preservation of species and the search for new energy sources. Not far in the background of all those new debates, however, are the oldest subjects of international politics—competition for resources and competing interests and ideas concerning economic growth, the distribution of wealth, and the terms of trade.

An important consequence of those developments is that the arenas in which environmental policy is determined are increasingly international—not just debates in the U.S. Congress, rulemaking proceedings at the Environmental Protection Agency, and implementation decisions by the states and municipalities, but opaque diplomatic "frameworks" and "protocols" hammered out in remote locales. To some, that constitutes a dangerous surrender of national sovereignty; to others, it heralds a new era of American cooperation with other nations that is propelled by the realities of an interdependent world. To policy makers themselves, it means that familiar questions of the benefits and costs of environmental rules are now enmeshed with questions of sovereignty and political legitimacy, of the possibility of large international income transfers, and of the relations of developed to developing countries.

In short, environmental issues are becoming as much a question of foreign policy as of domestic policy; indeed, the Clinton administration has made what it calls "environmental diplomacy" a centerpiece of this country's foreign policy.

AEI's project on global environmental policy includes contributions from scholars in many academic disciplines and features frequent lectures and seminars at the Institute's headquarters as well as this series of studies. We hope that the project will illuminate the many complex issues confronting those attempting to strike a balance between environmental quality and the other goals of industrialized and emerging economies.

CHRISTOPHER DEMUTH
IRWIN M. STELZER
American Enterprise Institute
for Public Policy Research

Irrational Elements of Environmental Policy Making

Irwin M. Stelzer

> Sir David Owen: *I do not quite know my-self what an economist is. What is his function? Is it to add to the cheerfulness and frivolity of life?*
>
> John Maynard Keynes: *My object, I think, in this was to promote clear thinking on these matters.*[1]

Discussions of environmental issues often generate more heat than light—and for good reason: these debates are about strongly held and widely different views of how the world works and how it should work. They are perhaps the most deeply ideological of any debate since the welfare state was established over the objections of those who believed in a much more limited central government. In this essay, I hope to demonstrate that debates about this or that rule, or this or that figure in some cost-benefit analysis, are so intense because the world views

---

1. "Statement before the Royal Commission on Lotteries and Betting, December 15, 1932," *The Collected Writings of John Maynard Keynes,* ed. Donald Moggridge, vol. 28 (London: Macmillan and Cambridge University Press, 1982), p. 408.

underlying what seem to be discussions of regulatory details are, in fact, a clash of what Tom Sowell calls "underlying assumptions about the world—a certain vision of reality."[2] These visions, adds Sowell, "compete with one another . . . for the allegiance of . . . a whole society."[3]

Arguments about environmental policy are, in reality, arguments between people with competing visions about fundamental issues. Despite the fact that most fair-minded people believe that environmentalism has made an important contribution to the quality of life at a cost reasonably commensurate with the benefits—although not at a cost as low as it might have been had greater reliance been placed on market-oriented solutions to environmental problems— a large gap exists between the world visions of environmentalists and those more skeptical of the need for further steps to improve our environment or to avoid future degradation. That gap is most clearly seen in three areas: attitudes toward economic growth (as economic growth is conventionally, but, alas, incorrectly measured),[4] attitudes toward the propriety of the distribution of income, and attitudes toward the appropriate extent of government control of individual behavior.

Economic Growth

The gap between those I will call "environmentalists" in this section of the essay—I will later distinguish between those for whom environmental values are absolute and

---

2. Thomas Sowell, *The Vision of the Anointed* (New York: Basic Books, 1995), p. ix.

3. Ibid.

4. The failure to include in reported gross domestic product and incomes such nonmarket outputs as environmental quality and leisure time muddies this argument; discussion of this issue would require a separate paper.

other, more nuanced thinkers—and those with a presumption in favor of economic growth is substantial. The latter associate economic growth with rising incomes,[5] improved living standards, an increased willingness and ability of the "haves" to share with the "have-nots," the ennobling impact of jobs for all, the spread of democratic institutions,[6] and even improved environmental quality.[7]

Not so, environmentalists. Economic growth, as they see it, places a strain on finite natural resources, in which are included everything from fossil fuels to clean air. The most famous modern-day statement of the position that "present growth trends" are heading us to disaster is, of course, the 1972 Club of Rome study.[8] The fear that we will run out of resources did not originate nor end with the Club of Rome, of course. Some one hundred years earlier William Stanley Jevons, styled by Joseph Schumpeter as "without any doubt one of the most genuinely original

---

5. See, for example, Adam Smith, *An Inquiry into the Nature and Causes of the Wealth of Nations,* ed. R. H. Campbell and A. S. Skinner, vol. 1 (Oxford: Clarendon University Press, 1976). Smith argues that countries experiencing a "continual increase" in national wealth pay the highest wages (p. 87); also, he writes, "Though the wealth of a country should be very great, yet if it has been long stationary, we must not expect to find the wages of labour very high in it" (p. 89).

6. Robert J. Barro, *Determinants of Economic Growth: A Cross-Country Empirical Study* (Cambridge, Mass.: MIT Press, 1997), p. 61: "Non-democratic places that experience substantial economic development tend to become more democratic."

7. "The real secret to environmental improvement is economic growth," according to Lee R. Raymond, chairman and CEO of Exxon Corporation. See his speech at the World Petroleum Congress in Beijing, October 13, 1997. See also Thomas G. Schelling, *Costs and Benefits of Greenhouse Gas Reduction* (Washington, D.C.: AEI Press, 1998). Schelling argues that the developing countries' "best defense against climate change . . . is their own continued development" (p. 4).

8. Donella H. Meadows, Dennis L. Meadows, Jørgen Randers, and William W. Behrens III, *The Limits to Growth* (New York: Universe Books, 1974).

economists who ever lived,"[9] worried about "our present rapid multiplication when brought into comparison with a fixed amount of material resources" and expressed the fear that Britain's industrial growth would come to a halt because its coal reserves were running out. Worse still: "It is useless to think of substituting any other kind of fuel for coal."[10]

Similar predictions of the impending exhaustion of the world's oil reserves crop up periodically and have typically been followed by the announcement of discoveries that exceed annual production, as a former scholar at Resources for the Future, Bruce C. Netschert, steadily argued when it was not fashionable to do so. And he has been proved right. When the Club of Rome report was published, total remaining reserves of crude oil amounted to 670 billion barrels. Since then, we have consumed 550 billion barrels. Yet reserves now stand at more than 1,000 billion barrels, 50 percent above the 1972 level. That, of course, does not deter such as John Avery, a professor at the University of Copenhagen, from worrying that an expanding and fuel-guzzling population "could burn all of the world's remaining reserves of fossil fuels in less than a century."[11]

The threat of resource exhaustion is not the only evil its critics see in economic growth. They argue that it also results in the production and accumulation of goods that people have been led to believe they need by insistent advertising,[12] rather than goods that satisfy "the basic mate-

---

9. Joseph Schumpeter, *A History of Economic Analysis* (Oxford: Oxford University Press, 1954), p. 26.

10. W. Stanley Jevons, *The Coal Question*, 3d ed. (1906), Reprints of Economic Classics (New York: Augustus Kelley, 1965), pp. 184, 454.

11. John Avery, *Progress, Poverty, and Population: Re-reading Condorcet, Godwin and Malthus* (London: Frank Cass, 1997), p. xiii. But see Gertrude Himmelfarb, "The Ghost of Parson Malthus," review of *Progress, Poverty, and Population: Re-reading Condorcet, Godwin and Malthus,* by John Avery, *Times Literary Supplement,* January 23, 1998, p. 4.

12. "With a few notable exceptions, the most powerful influences on

rial needs of each person on earth";[13] forces us to rely increasingly on fossil fuels that foul the environment and threaten us with the catastrophe of global warming;[14] and makes us dangerously dependent on oil imports from unstable parts of the world.[15]

Consider, as one example of a reflexive antipathy to growth, then Senator Al Gore's description of the construction of a large housing and shopping development in Virginia, one that would permit many families to realize the American dream of home ownership, create jobs, and make life and shopping a bit easier. This construction is, to Mr. Gore, another example of "human kind's assault on the earth. . . . As the woods fell to make way for more concrete, more buildings, parking lots, and streets, the wild things that lived there were forced to flee. Most of the deer were hit by cars."[16] No benefits of economic growth leapt to Gore's mind as he drove through thriving Virginia: only costs.

---

popular attitudes in upper-income countries—advertising and entertainment—promote over-consumption and waste," *Caring for the Earth: A Strategy for Sustainable Living* (Gland, Switzerland: World Conservation Union, United Nations Environment Programme and the World Wide Fund for Nature, October 1991), p. 520, hereinafter cited as *Caring for the Earth.* For a fuller discussion of advertising and its critics, see John E. Calfee, *Fear of Persuasion: A New Perspective on Advertising and Regulation* (Washington, D.C.: AEI Press, 1997).

13. *Caring for the Earth,* p. 24.

14. Our burning of fossil fuels "amounts to an addiction" that is "bringing global environmental catastrophe." The environmental ills consequent upon the use of fossil fuels "are killing our water, our air, our plants, our animals, and eventually, if not checked, they will kill us," according to Senator George J. Mitchell, *World on Fire: Saving an Endangered Earth* (New York: Macmillan Publishing Company, 1991), p. 47.

15. On this latter point see, for example, Alliance To Save Energy et al., *America's Energy Choices: Investing in a Strong Economy and a Clean Environment* (Cambridge: Union of Concerned Scientists, 1991), *passim.*

16. Al Gore, *Earth in the Balance: Ecology and the Human Spirit* (Boston: Houghton Mifflin Company, 1992), p. 25.

This hostility to economic growth, and to the accompanying increase in material affluence, permeates much, although not all, environmentalist thinking. Consider the views of a prominent trio of environmental organizations. Convinced that "economies and societies different from most that prevail today are needed if we are to care for the Earth [note: *earth* is often capitalized in environmental literature] and build a better quality of life for all,"[17] these organizations contend that "affluence has not protected high-income countries or the wealthy minority in poor countries from drugs, alcoholism, AIDS, street violence and family breakdown."[18] The fact that poverty has certainly not done very much to eliminate these scourges remains unnoted, as have the facts that their incidence is considerably higher among the poor than among the affluent and that it is only the so-called high-income countries that can afford to take effective measures against these ills. Rich remains better than poor, as every poor country knows.

This antigrowth bias of many environmentalists was, and to an extent that few will admit still is, matched by a progrowth bias on the part of their opponents. Industry leaders, many of them remembering the days before World War II when factories stood idle and the unemployed lined up at soup kitchens, thought it their duty to operate at full throttle and to satisfy the desires of consumers for more and better products. To them, and to their successors, economic growth not only meant higher profits but was, and remains, every bit as compelling a moral imperative as preservation of the environment was to the newly minted greens. To grow is to provide jobs, houses, and the material things of life to consumers, dividends to investors, and national strength and prestige to America. There can be no higher calling.

---

17. *Caring for the Earth*, p. 1.
18. Ibid., p. 20.

Many industrialists quite correctly saw early efforts by environmentalists to persuade or, worse still, to force them to consider the environmental impacts of their production practices as an assault on these values, as a threat to their management prerogatives and to the profitability of their companies, and as part of an effort to undermine free markets and, indeed, capitalism itself. Their fears were fanned by the fact that early-day environmentalists often made common cause with opponents of capitalism and with other radical sorts with whom denizens of the nation's board rooms did not communicate very well. Although environmentalists are no longer seen as anticapitalist flamethrowers, many businessmen still find them threatening. They fear that the new breed of greens is engaged in a relentless drive to pile regulation upon regulation, with no attention—indeed, a studied inattention—to a comparison of costs and benefits.

In the case of global climate change, the suspicions of industry are heightened by the fact that many in favor of strict control of greenhouse gas emissions favor carbon taxes as the instrument to achieve their goal. Here, too, are agendas within agendas. The business community knows that antitax sentiment runs so high among the electorate in most countries that it is not possible for politicians to support increases in income or sales taxes. Starved of additional sources of revenue, governments cannot expand the scope of their activities—just the situation the generally conservative business community favors. Carbon taxes, however, represent what seems to be a politically acceptable new source of revenue, one that can fuel further expansion of the welfare state. So the battle over global climate change, and over "green" taxes, becomes a battle between advocates of a restricted and advocates of an expansive role for government in the next century—a situation designed to make the battle more intense than it would otherwise be.

Industry's rallying point in the battle is cost-benefit analysis—restrict any new environmental programs to those the benefits of which exceed the costs. But this devotion to cost-benefit analysis by most of industry comes late. Earlier in the debates, environmentalists—pressing the quite sensible case that there might be certain external costs associated with the production techniques of modern industrial societies and that methods should be developed to internalize these costs—were not given much of a hearing. In part, the willingness of environmentalists to jump from a demonstration that externalities exist to a call for regulation, or taxes, or both explains the reluctance of industry to credit their views. Businessmen intuit what Coase has shown: that a leap from identifying externalities to demanding taxes and regulations is a leap too far and, in the absence of high transactions or transition costs, is most likely to be counterproductive.[19] But in part the unwillingness to give environmentalists a hearing stems from the quasi-religious fervor that many executives display when discussing economic growth. It was not so long ago that business executives equated the smoke from factory stacks with prosperity—a charming anachronism, but an anachronism nonetheless. They did not acknowledge the fact that their cries of laissez faire had for centuries been raised "as a defense of . . . vested interests who were imposing important external costs on society by unsanitary working and living conditions, child labor, pollution, etc."[20]

Of course, there is little overt anticapitalist rhetoric emitting from today's environmentalists and little overt green-bashing or pleas to be allowed to impose external costs on society in the public statements of today's public-relations–smoothed "corprocrats." Political effectiveness

---

19. R. H. Coase, "The Problem of Social Cost," *Journal of Law and Economics*, vol. 3 (October 1960), pp. 1–44, conveniently reprinted in *The Firm, the Market and the Law* (Chicago: University of Chicago Press, 1988), pp. 95-156.

20. Thomas Sowell, *Classical Economics Reconsidered* (Princeton: Princeton University Press, 1974), p. 30.

goes not to the openly radical on either side but to those who speak the language of compromise and concern.

Besides, America's largest corporations are not insensitive to the fact that they are now well positioned to game the regulatory process, to the disadvantage of their smaller competitors and of potential new entrants. Anacharsis, a sixth-century Scythian prince, with what the *Oxford Classical Dictionary* describes as "a high reputation for wisdom,"[21] is believed to be the source of an observation cited by Jacob Viner: "Laws are merely spider webs, which the birds, being larger, break through with ease, while the flies are caught fast."[22] Had poor Anacharsis not been executed for his religious beliefs, he might have become a successful environmental lobbyist for some Fortune 500 company, devising regulations that would surely ensnare tiny flies, leaving his client-birds free to fly relatively unimpeded.

But we should not be misled by the more politic rhetoric of the combatants in the environmental policy arena: in the tussles that lie ahead, the old antigrowth and progrowth prejudices still lurk in the hearts of men, to paraphrase an old radio program. To Al Gore's acolytes, more remains less. And to many of America's industrialists, the notion of interfering with full employment, international competitiveness, and the onward march of material progress seems a form of sacrilege, especially in a nation that is already spending $144 billion annually to comply with environmental regulations.[23]

21. Simon Hornblower and Antony Spawforth, eds., *The Oxford Classical Dictionary*, 3d ed. (Oxford: Oxford University Press, 1996), p. 79.

22. See Jacob Viner, *Essays on the Intellectual History of Economics*, ed. Douglas A. Irwin (Princeton: Princeton University Press, 1991), p. 289. Viner says that these words "are attributed to Anacharsis."

23. Office of Management and Budget, Office of Information and Regulatory Affairs, *Report to Congress on the Costs and Benefits of Federal Regulations*, September 30, 1997, p. 29. These are 1996 dollars. The benefits are estimated at $162 billion (p. 33). Both figures are the subject of some controversy.

Income Distribution

Competing views of the desirability of economic growth and higher material standards of living are not the only deeply held visions that lie beneath the surface of the debate over environmental policy. So, too, do views about the propriety of the income distribution system in this country and, indeed, in the world. The relatively recent contention by the Environmental Justice Movement that environmental degradation in America is concentrated in low-income and minority neighborhoods—that costs go to the poor, benefits to the rich—is only the latest manifestation of the distributionist argument.[24]

On the international level, green groups argue that the industrialized world in general, and America in particular, consumes a disproportionate amount of the world's resources and produces a disproportionate amount of its pollution. This contention usually takes the form of charging the United States with consuming more oil and producing more pollution per capita than other countries,[25] as if some concept of fairness dictates that each person in the world is entitled to an equal number of gallons of gasoline or as if equal per capita shares were economically optimal. This egalitarianism run riot ignores the fact that America's energy-intensive agriculture feeds most of the world, while populations reliant on labor-intensive agricultural techniques often starve; that America's huge distances compel the use of extensive trucking systems to distribute

---

24. President Clinton, by executive order, established environmental justice as "a national priority." See Environmental Protection Agency, "Environmental Justice," http://es.epa.gov/oeca/oejbut.html, November 6, 1997. See also the president's memorandum for the heads of all departments and agencies, http://www.epa.gov/docs/oejpubs/prezmemo.txt.html, February 11, 1994.

25. See, for example, National Academy of Sciences et al., *Policy Implications of Greenhouse Warming* (Washington, D.C.: National Academy Press, 1991), p. 7.

goods; and that this country's share of the world's output of CO_2, although disproportionately large relative to its share of the world's population, is in line with its share of the world's output of goods and services.

The redistributionist underpinnings of the environmental movement make discussions of genuinely international problems extraordinarily difficult because the solutions proposed by self-styled environmentalists are aimed at twin goals: protecting the environment and redistributing the world's income. Thus, the World Wildlife Fund for Nature uses its web site to encourage debate on the following topic: "The rich must live more simply, so that the poor may simply live" and on the question of whether sustainable development includes among its prerequisites "the need for income redistribution."[26]

The environmentalists' dissatisfaction with the way the world's income is distributed and its resources consumed comes face to face with the conviction on the part of many in American industry that incomes quite properly reflect marginal revenue product, that the earnings of working people (including themselves) are proportionate to their contributions to society, and that programs that aim to take from the haves and give to the have-nots are likely to have unfortunate consequences, not least among them a reduction in the incentive of the world's most productive members to produce at full bore.

Thus, calls to reduce Americans' consumption of resources so that more will somehow become available to poorer nations, which redistribution is what much of the new international environmental movement is all about, raise the hackles of those satisfied with the current distribution of income. Those efforts also worry economists who fear that any disjunction between contribution to output and the reward for that contribution will inefficiently reduce total welfare.

---

26. "Sustainable Development," http://www.panda.org/resources/factsheets/enviroecon/03sustdev.htm, 1993.

It should be noted that defenders of the current system of income distribution do not always come to the debate with clean hands, for their own incomes are often being determined by a cozy arrangement with members of corporate compensation committees not famous for an unalloyed devotion to equating executive compensation with marginal revenue product.

The two issues under discussion—economic growth and income distribution—are not unrelated. Rapid growth has historically been accompanied by rising living standards for all groups, making economic growth perhaps the most successful of all antipoverty programs. As William Baumol and his colleagues point out, "Rising incomes and the fruits of the technological revolution have filled our lives with goods and services unavailable, and even unimaginable, 100 years ago and, perhaps most important, the revolution has produced its most dramatic changes in the lives of the millions of ordinary working people." [27] Indeed, although the benefits of this growth in the economy and in productivity were not distributed evenly, it is a fair conclusion that all income classes benefited: "Even welfare recipients today are hardly expected to subsist on . . . one bowl of gruel . . . [and] the perpetual threat of famine . . . has disappeared in this country and other industrialized lands. The end of that spectre is economic progress indeed, even for the poorest members of the community."[28]

The conviction that economic growth, as we measure it, benefits all groups makes it difficult for those who hold it to find common ground with proponents of programs designed to reduce or even halt that growth in the interests of avoiding environmental degradation.

---

27. William J. Baumol, Sue Anne Batey Backman, and Edward N. Wolff, *Productivity and American Leadership: The Long View* (Cambridge, Mass.: MIT Press, 1989), p. 57.

28. Ibid., pp. 58–59.

Government Control of Individual Behavior

The final factor making the environmental debate so heated is that it is merely a subset of a broader debate over the extent to which the government should be asked or allowed to interfere in the lives of its citizens. Many see regulations that require them to wear seat belts in their very own cars; not to smoke in their very own offices; to choose smaller, more dangerous over larger, safer cars; to suffer the indignities of low-flow toilets; and not to keep guns in their very own homes as infringements on their personal liberty. Especially offensive are restrictions on the use or design of the automobile, which machine Americans quite properly see as providing the personal mobility that totalitarian regimes so fear.[29] Add urgings by assorted bureaucrats not to eat this or that food, or to boycott this or that toy that some child might decide to chew rather than cuddle, and you have in the minds of some a "nanny state." Never mind even the most closely reasoned arguments about external costs and market failure: outside the Washington Beltway, freedom takes precedence.

Others contend that it is proper for government to look after the health of its citizens, if for no other reason than that some of the cost of unhealthful behavior will fall on the public finances. What to others seems an infringement on individual choice is to regulatory activists the use of the resources and wisdom of government to channel people into the proper cars—or better still, onto mass transit—or at least into high-occupancy–vehicle lanes on their highways and into health food aisles in their supermarkets.

---

29. James D. Johnson, *Driving America: Your Car, Your Government, Your Choice* (Washington, D.C.: AEI Press, 1997), *passim.* See also George F. Will, *The Woven Figure: Conservatism and America's Fabric* (New York: Scribner, 1997), pp. 358–59: "An open road produces an open society. The automobile has been an emancipating device. . . . Were Huck [Finn] to light out for the territories today, he would go in a Ford Explorer."

This conflict of visions comes to the fore as solutions to the alleged problem of global warming are mooted. As Linda Stuntz, a particularly thoughtful observer of the energy scene and a combatant in the environmental wars by virtue of her prior positions in government and her current representation of a large coal-burning utility, recently pointed out, "At least some adherents of the Politically Correct Energy Future . . . are ready to expand the reach of government to impose unprecedented restrictions on our freedom. We all accept restrictions on our freedom every day to benefit the common good. At some point, however, we need to ask ourselves whether these restrictions are truly worth the infringement on our liberty."[30] And when we do, we can be certain that we will get wildly different answers from believers in the necessity of government intervention and from those more concerned with individual freedom of choice and action.

Interestingly, we see here a strange confluence of the views of activist environmentalists and social conservatives. The former *know* that it is better for us to ride on bicycles or use mass transit than to tool around in giant air-conditioned, gas-guzzling sports utility vehicles. They *know*, too, just how much glass an architect should be allowed to design into a house (the goal being to minimize fuel consumption for heating and cooling) and how many people should be allowed to visit our national parks. Social conservatives, supposedly at the opposite end of the political spectrum from the environmentalists, whom they tend to view as wreckers of the American economy, *know* which television programs we should be allowed to watch and which should be banned. I gather that old movies are "in" but new sexy ones are "out"; that Poirot is fine but that *The Simpsons* is a threat to family values. (I believe, but am not

---

30. "Global Warming and Energy Policy: Separating Fact from Fantasy," presentation to the Center for Energy and Economic Development's annual board meeting, November 7, 1997, p. 6 (mimeo).

certain, that it was Irving Kristol who first pointed out that liberals are opposed to showing violence on television, while conservatives are opposed to televised sex.)

In short, both environmentalists and social conservatives would substitute regulation—or, as a second choice, taxes in the case of environmentalists and tax credits in the case of conservatives—for the market. Environmentalists have recently mounted an attack on the public's preference for big (safe) cars; social conservatives, or at least those not content with calling for voluntary, private sector boycotts, continue to attempt to control television fare and Internet access. Little wonder that our vice president finds both groups to his liking, granting the one all it desires by way of environmental restrictions and the other approval for its efforts to dictate programming (more for kids, less for grownups), impose a government rating system on television networks, and supervise a "voluntary" program of controlled access to the Internet. We can only hope that he never seeks to satisfy both groups simultaneously, lest we be reduced to getting to the cineplex on our bicycles to find that we can see only *Mary Poppins* and *The Sound of Music*, while munching on unbuttered and unsalted popcorn.

The desire to limit individual choice, of course, is not a new phenomenon. And it is closely related to the notion that a steady increase in the accumulation of material goods is somehow bad for us and also threatens social harmony. This view has its roots in the sumptuary legislation that made its appearance as early as 594 B.C. Those laws varied over time but had in common restrictions on private consumption to prevent extravagance, preserve class distinctions in dress and entertainment, and preserve public morality.[31]

---

31. Sumptuary legislation was based on "the feeling that luxury and extravagance were in themselves wicked and harmful to the morals of the people," Frances Elizabeth Baldwin, *Sumptuary Legislation and Personal Regulations in England* (Baltimore: Johns Hopkins University Press, 1925), p. 10.

The assumption of these laws is that less is more and that the state knows what is best for its citizens—how many gold threads a cloth may contain or how many chickens may be consumed at a feast. These laws have their modern counterparts. One scholar, writing as early as 1934—some sixty years before Hillary Clinton assumed responsibility for the nation's health—observed, "Laws restraining and forbidding the use of liquor and tobacco have . . . kinship with sumptuary legislation, for they are based upon the same principle, the protection of individual and public welfare and morality."[32]

And before Hillary we had Jimmy—Carter, that is. He told us that "too many of us now tend to worship self-indulgence and consumption. . . . But . . . owning things and consuming things does not satisfy our longing for meaning . . . [and] piling up material goods . . . cannot fill the emptiness of lives which have no confidence or purpose."[33]

From attitudes such as President Carter's and Mrs. Clinton's, it is an easy step to laws that tell us how fast we may drive, how cool and warm we may keep ourselves, how and of what materials we may build our houses, whether we may have frost-free refrigerators and self-cleaning ovens, whether we must join a car pool to have access to all the lanes of our tax-financed highways, how much water we may use to flush our toilets, and what fuels we may use to keep our lights on and our factories running.

The environmentalists are not alone in their desire to impose their views of the good life on others. Many of the usual crowd of conservative critics have their own notions of that good life and of socially acceptable behavior and are equally willing to impose them on others. It is difficult to defend as God-given the right to drive a sports utility

---

32. J. M. Vincent, "Sumptuary Legislation," *Encyclopaedia of the Social Sciences,* vol. 14 (New York: Macmillan Group, 1934), p. 466.

33. Jimmy Carter, "The Crisis of Confidence," address to the nation, July 15, 1979, p. 4 (mimeo).

vehicle while puffing on a cigarette at the same time denying that people have a right to watch whatever television program suits their fancy.

That leaves three groups that stand against the "green machine" of regulators and taxers: industry, free market economists (who, of course, recognize the existence of external costs and market failure), and a cadre of multidisciplinary researchers dispassionately searching the data for guides to sensible policies—a group represented by the author of the other essay in this volume.[34]

Industrialists have some difficulty being heard because they are perceived to be acting out of self-interest and because the more powerful of them speak in muted tones,[35] both because they do not wish to alienate a possible future president and because they have confidence that they can survive any regulation aimed equally at their competitors.

That leaves two groups eligible for the battle: free market economists and multidisciplinary research teams rendered dispassionate by the benign environment of a well-ordered think tank. The latter, of course, include economists, and so these groups overlap. That view will be well represented in the accompanying essay. I will speak

---

34. Libertarians might be counted a fourth group. But in this case they seem to be eligible for inclusion among the free market economists, with whom they seem prepared to make common cause. Witness the following from one of that group's intellectual leaders: "The protection of the clean air has become an authentic public good, and under these circumstances [multiple polluters affecting multiple property owners] government legitimately acts as a forum for deciding how clean is clean enough and for crafting legislation to produce the desired result," Charles Murray, *What It Means to Be a Libertarian* (New York: Broadway Books, 1997), p. 115.

35. There are honorable exceptions. See, for example, Lee Raymond's comments at the World Petroleum Congress on the "vast international bureaucracy responsible to no one" that Raymond contends would be required, along with "punishing, high energy taxes," to achieve the reduction in carbon dioxide emissions that global warming advocates are seeking.

only for free market economists, with their kit of tools—prices, discount rates, costs, and cost-benefit analysis. The best of the breed recognizes two things: that markets are not perfect and that their tools are not value free. They also assume that it is worth the effort to quantify such benefits as the value of a hike in a national park and such costs as those associated with forgoing or sharply curtailing the use of fossil fuels. The more sensible keep in mind historian Eric Hobsbawm's observation that "from time to time history catches economists at their brilliant gymnastics and walks off with their overcoats"[36] and so recognize that their efforts at quantification are not precise but are best aimed at helping policy makers avoid the most egregious of the available errors.

But neither the fact that free market economics represents a system of values nor that a bit of humility is appropriate in applying the quantitative tools of the economist's trade can be taken to mean that economists have little to offer in the debate over appropriate environmental policy. Indeed, no other group can contribute as much. Environmentalists who decry the use of cost-benefit analysis are essentially asking for an open ticket to impose their vision of a good society on everyone else and for a blank check with which to pay for that imposition.

For economists to be effective in promoting clear thinking, to recall the phrase used by John Maynard Keynes cited at the beginning of this essay, they must first recognize that they are in a battle with opponents whose tactics have been beautifully described by Sowell. The key elements of these tactics are:

1. Assertions of a great danger to the whole society . . . .
2. An urgent need for action to avert impending catastrophe.

---

36. Eric Hobsbawm, *On History* (New York: New Press, 1997), p. 95.

3. A need for government to drastically curtail the dangerous behavior of the many, in response to the prescient conclusions of the few.
4. A disdainful dismissal of arguments to the contrary as either uninformed, or motivated by unworthy purposes.[37]

To many economists who inhabit campuses on which it invites ridicule to challenge the vision of the anointed, these tactics have proved intimidating: better to build huge, meaningless macroeconomic forecasting models that do not antagonize colleagues, but do titillate the media and impress the tenure committee, than to wrestle with controversial microeconomic issues,[38] particularly if one is of the view that the academic prejudice in favor of drastic government action (and more grants) is wrong.

For those who choose to do battle in the environmental arena, I can only advise the use of an assortment of weapons: a sense of humor when attacked as a philistine for defending the virtues of economic growth and rising material well-being; a sense of humility when offering conclusions based on quantitative assessments of difficult-to-measure phenomena; a recognition of the fact that market failure can occur and does warrant government action, even given the risk of regulatory failure; a realization that the costs and benefits of policy changes should be measured by comparing not some ideal circumstance but the real-world present situation with the aimed-for new situation; an understanding that economics can inform policy judgments but need not always supersede noneconomic con-

---

37. Sowell, *The Vision of the Anointed*, p. 5.
38. Why economists waste their time with forecasting is a mystery. "The fact is economists are poor at forecasting short-term changes in the economy. . . . Microeconomics attracts less attention from the media than does macroeconomics, yet it has had stunning practical successes during the past decade," Gary S. and Guity Nashat Becker, *The Economics of Life* (New York: McGraw Hill, 1997), p. 311.

siderations;[39] and the conviction that rejection of the use of economic analysis will impoverish us materially and deprive us of an anchor when the gales of special privilege hit the ship of state.

39. Armatya Sen is not alone in reminding us that "modern economics [is] largely . . . an offshoot of ethics. . . . The ethics-related tradition goes back at least to Aristotle, " *On Ethics and Economics* (Oxford: Basil Blackwell, 1987), p. 3.

A Rationalist Program for Environmental Policy Making

Paul R. Portney

his essay sets out a model of environmental decision making in what I call the rationalist mode. I
shall first lay out the three elements of the rationalist mode of environmental decision making; then talk about
the progress, or lack thereof, over the past two decades in
introducing this approach into environmental decision
making; and finally turn to problems with each of the three
elements of the rationalist mode.

Balancing Benefits and Costs

The first element, or principle, is that we ought to balance
the incremental benefits of a proposed policy change with
incremental costs at the margin. This idea might be
summed up as follows: "Wisdom requires balancing the
good against the bad, to determine whether the overall
result is positive or negative." Interestingly, this quotation
is found on page 188 of Vice President Al Gore's book,
Earth in the Balance, and it stands in stark contrast to the
very critical remarks about economics contained elsewhere
in the book.

This first principle, then, has to do with the desirability
of balancing, qualitatively or quantitatively, the good against
the bad that any proposed policy intervention will do. To

do that, we rely on studies of the costs and benefits of environmental regulatory proposals. On an aggregate basis, studies show that the total cost, in 1997 dollars, of complying with all federal air, water, solid waste, hazardous waste, pesticide, toxic substance, and drinking water regulations is somewhere between $100 billion and $160 billion per year. Paradoxically, the lower-bound estimate comes from the Bureau of Economic Analysis of the Department of Commerce and the high estimate from the Environmental Protection Agency.

More recently, the Office of Management and Budget made its own estimates of these environmental compliance costs. Its figure of $144 billion in 1996 is well within the range bounded on the lower end by the Commerce Department and on the upper end by the Environmental Protection Agency. If we take $150 billion as a reasonable estimate, we see that we are spending 2.2 percent of our gross domestic product on environmental compliance. Although it is difficult to compare this figure with the level of the commitments to environmental compliance of other developed countries—the data from those countries are very poor—I think it safe to conclude that America devotes significantly more to environmental protection than our nearest competitors. The Netherlands and Germany, for example, appear to spend about 1.8 percent of their GDP on complying with their environmental regulations. Therefore, the charge often heard both before and after the Kyoto conference that the United States is not doing its share of environmental compliance is not supported by the data (although it is true that energy prices are much lower in the United States than anywhere in Europe).

So much for aggregate data. As I mentioned earlier, the relevant comparison for policy making is of the incremental benefits and incremental costs of individual regulations. The two rulemakings receiving most attention of late are EPA's recent tightening of the ambient air quality standard for ozone and its announced new fine particle standard. The annual incremental cost of these additions to environmental regulation is likely on the order of $10 billion per year (again subject to considerable uncertainty). This is not, as some people have suggested, the end of economic life as we know it, but there is no question that the marginal cost of meeting these tighter new standards will be quite significant, once they are fully implemented.

So, too, with climate change. The international negotiations in Kyoto in December 1997 resulted in an agreement under which the industrialized nations are to reduce their emissions of carbon dioxide and five other greenhouse gases by an average of 5 percent below 1990 levels by no later than the year 2012. Under the protocol, the United States is to reduce its emissions by slightly more—about 7 percent. It is still much too soon to know whether the policies will be put forward, much less implemented, that would be required to meet this ambitious goal. If this goal were to be rigorously pursued (the smart money says it will not, at least for the time being), it would prove to be quite costly. While such estimates are very uncertain, and may prove to be too high, it currently appears that it would cost the United States $20–40 billion annually to meet that goal. Again, however, the United States is unlikely to make an all-out effort to reduce its emissions by the requisite amounts at the required deadlines, especially if the Senate fails to ratify the climate treaty.

So much for the cost side of the equation. The benefits of environmental regulations are generally more difficult to pin down. When they are, however, they are sometimes found to be well in excess of the costs. Virtually every analysis of the 1970 amendments to the Clean Air Act, for example, suggests that the benefits associated with those changes— although they could have been had much less expensively than they were—are well above the costs, possibly by an order of magnitude or more.

It is the case, of course, that since 1970, when the EPA was created, we have been addressing the most obvious and distressing threats to health and the environment. Today, after almost three decades of regulation, the problems that remain to be solved quite naturally offer smaller marginal benefits. The benefits of reducing trichloroethylene concentrations in drinking water from four parts per billion to three, for example, are probably very small.

That is in contrast with many environmental regulations that were written in the early 1970s, following a time when the Cayahoga River had spontaneously combusted and when air pollution was so bad in Pittsburgh that at noon on an August day, authorities had to turn on the street lights and drivers had to turn on their headlights. The benefits of regulations at that point were obviously quite considerable.

More recent changes in laws and regulations necessarily strike a finer (and sometimes unfavorable) balance between benefits and costs. We are, it is important to note, doing a better job of striking this balance. The changes made in 1996 to the Safe Drinking Water Act and the elimination of the Delaney clause and its replacement with the Food Quality and Protection Act, for instance, represented progress. For the first time, statutes that in the past had prohibited federal regulators from striking qualitative balances between benefits and costs were replaced with laws giving them the power to do so and directing them to attempt an appropriate balance.

Least-Cost Approaches

The second pillar of the rationalist approach is that we try to achieve our goals using least-cost approaches. It is comforting to note that the rhetoric on this score is much better today than it has ever been in the past, although I emphasize the word *rhetoric*. Gone are the days when charges of "licenses to pollute" were rife. Now those active in both the environmental advocacy and the regulatory communities talk approvingly of and are concerned about developing least-cost approaches. And they recognize that taxes on pollution, marketable permits, and deposit refund schemes—to name but a few such approaches—are useful weapons in the environmental-regulatory arsenal.

Triumphs in this area are not confined to rhetoric. There are also triumphs in real-world policy making. These victories began with the lead phase-down in gasoline during the 1980s. Due to the efforts of, among others, Robert Hahn (now at the American Enterprise Institute), a market was created for the lead that was used in gasoline refining in the 1980s. That successful experiment led to the trading program for sulfur dioxide emission reductions that was instituted in Title IV of the 1990 amendments to the Clean Air Act, marking the first time that an ambitious economic incentive-based approach to environmental regulation was enshrined in federal legislation.

This was no small victory. Had we taken the old, forced-scrubbing approach to reducing sulfur dioxide emissions from coal-fired power plants—the staple of federal environmental policy until 1990—we would be spending, by my estimation, $5–7 billion annually to obtain the 8.9 million ton reduction in SO_2 that was legislated that year. As a result of using flexible trading approaches, giving affected power plants the right to decide how they will meet reduced emissions—by switching to natural gas, by cleaning coal in advance, and the like—we are spending only about $2 billion per year to obtain that same 8.9 million ton reduction in SO_2. A policy that allows us to meet a given environmental objective, and save $3–5 billion per year in the process, must be counted a triumph.

Similarly, one of the components of the Clinton administration's policy on global climate change is an insistence on a role for tradable permits, giving affected sources, wherever they may be, the opportunity to find least-cost, or lesser-cost, ways of reducing CO_2 emissions any place in the world.

We have thus made substantial progress, both rhetorically and substantively, in environmental policy, by observing the second principle of the rationalist approach. But we still have a long way to go. Unfortunately, for instance, the other two major components of the 1990 amendments to the Clean Air Act, those dealing with urban air quality and so-called hazardous air pollutants, were premised on overly expensive, old-fashioned command and control. In short, we got it one-third right with respect to acid rain but two-thirds wrong when dealing with traditional air pollution problems and hazardous air pollutants: we missed opportunities to expand the market-based approach to the control of those pollutants.

Environmental Federalism

Applying the third element of the rationalist approach, we consider the possibility that not every problem ought to be handled at the federal level. It seems sensible to me that we engage in some creative environmental federalism. Why not invest the power not merely to implement and enforce regulatory programs but to set the actual standards at the appropriate level of government?

Unfortunately, this tenet of the rationalist approach has not been reflected to any significant degree in recent environmental policy. One might have expected some progress with the election in 1994 of the Republican Congress. After all, in the campaign leading up to those elections, there was much talk of moving authority for problems that are not really federal problems to lower levels of government. Yet, there has been no progress since the 1994 elections, not even a constructive debate in Congress about the appropriate locus for environmental decision making.

Thus, even though the 1996 amendments to the Safe Drinking Water Act incorporated one element of what I have here called the rationalist mode of policy making (introducing a rough balancing of benefits and costs), those amendments did not feature a discussion about even the *possibility* that standards for drinking water contaminants might best be set not in Washington, under a one-size-fits-all approach, but by state governments. After all, there is no interstate spillover in the case of drinking water, as there is with air pollution, water pollution, and some other environmental problems.

Still, progress, although slight, is being made. For instance, the Environmental Council of the States, an association of the heads of the state environmental regulatory agencies, played a very influential role in the recent Ozone Transport Assessment Group deliberations, deciding how much of the blame for ozone problems in one state should be apportioned to pollution that originates in another state. Notably, EPA Administrator Carol Browner encouraged these discussions. The hope is that while some pollution control efforts will always be appropriately the province of the federal government (air and water pollution being the obvious examples), we can at least debate whether drinking water and solid and hazardous waste might best be left to lower levels of government.

Problems with the Rationalist Approach

Let me now turn to the problems with my proposed rationalist approach. First, it may place too great a faith in the precision of cost-benefit analysis. We need a substantial dose of humility when discussing this technique. One of the reasons that we have made less progress in the legislative arena than we otherwise might is that some economists have overstated what benefit-cost analysis can deliver by way of assistance in standard setting. Benefit-cost analysis was never conceived of as providing a decision rule. Rather, it is a decision tool that can help in establishing standards at the same time we take other considerations into account.

Neither is benefit-cost analysis value free; yet, some proponents embrace it with the same semireligious fervor that environmentalists often bring to discussion of natural resource use or pollution policy. To offer but one example, benefit-cost analysis uses market prices for labor, capital, and natural resources as a measure of the opportunity cost of those resources. But those prices depend critically on the distribution of income in our society, a distribution that many challenge on both economic and social grounds.

In sum, we need to recognize the shortcomings of benefit-cost analysis and be open about them, if we expect people to continue to rely on this tool in making environmental decisions. Benefit-cost analysis is very difficult to apply in practice, the uncertainties are great, and complicated issues such as selecting a discount rate abound.

The second pillar of rationalist decision making, least-cost approaches, is also not without weaknesses. Simply put, incentive-based alternatives to command-and-control requirements are more difficult to design, and particularly to implement, than economists have acknowledged. In some sense, we have become seduced by the ease with which we established a permit system for trading SO_2 emissions to reduce acid deposition in the United States. It would be wrong to draw a direct analogy to climate policy from our experience with acid rain. Putting in place a well-functioning permit market for CO_2 and other greenhouse gases will prove a much more difficult problem to solve.

Consider the attempts of Southern Californians to institute an air pollution permit trading system. Because emissions in one part of the Los Angeles basin contribute more to ozone formation than emissions in other places and because emissions in the morning and early afternoon hours create more smog than late afternoon and nocturnal emissions, authorities are considering the creation of a spatially and temporally differentiated permit system. This becomes extraordinarily complicated, and academics with the intellectual ability to design conceptually pure systems often do not have the patience to sit with legislators, regulators, and their aides for months and years at a time to translate their research into legislative language. Fortunately for Southern Californians, John Ledyard, Glen Cass, and their colleagues at Cal Tech have been an exception to this rule. But, difficult though it may be to institute incentive-based approaches, we must make the effort. We spend $150 billion per year on environmental compliance, so a saving of just 10 percent of that amount more than qualifies for what the late Senator Everett Dirksen referred to as "real money."

The third pillar of the rationalist approach to environmental policy, thinking creatively about the right level of government in which to rest power, faces greater resistance than the expanded use of either benefit-cost analysis or incentive-based environmental programs. Suggestions for taking control of some standard-setting authority away from the federal government and devolving it to governors, or even to mayors, evokes images of a "race to the bottom," a competition for environmental degradation. It is as if opponents of state-level standard setting believe that ordinary citizens (whom these centralists claim are virtually unanimous in their support for regulation) will not bounce out of office any governor sacrificing their health on the altar of economic growth. There is worse: even in a Republican Congress, the notion that a committee or subcommittee chairman will surrender jurisdiction over an important environmental statute, which he or she can use to raise funds during the election cycle, to a state governor or some lower-level official, makes the political economy of defederalizing environmental statutes most difficult.

To summarize, then, we can point with pride to progress made in enlarging the role of the rationalist mode in environmental policy making. This is particularly so with the use of benefit-cost analysis and incentive-based approaches. Even in these two areas, however, not to mention the third pillar of environmental federalism, there is much to be done. This work will be best done if proponents keep firmly fixed in their minds the fact that these three pillars are no more panaceas than is the current command-and-control apparatus they seek to dismantle. After all, no approach is.

About the Authors

PAUL R. PORTNEY is president of Resources for the Future. He received his B.A. in economics and mathematics from Alma College (Michigan) and his Ph.D. in economics from Northwestern University. He has been a visiting professor at the Graduate School of Public Policy at the University of California at Berkeley and a visiting lecturer at Princeton University's Woodrow Wilson School. In 1979–1980, he was chief economist at the Council on Environmental Quality in the Executive Office of the President. Mr. Portney has served on the Board of Environmental Studies and Toxicology of the National Academy of Sciences and on the National Oceanic and Atmospheric Administration's Panel on Contingent Valuation. From 1994 to 1997, he was a member of the Executive Committee of the Science Advisory Board of the Environmental Protection Agency and chairman of the board's Environmental Economics Advisory Committee. He has published widely on the costs and benefits of environmental regulation, including the forthcoming second edition of *Public Policies for Environmental Protection*.

IRWIN M. STELZER received his Ph.D. degree from Cornell University and his bachelor's and master's degrees from New York University. He has been director of the Energy and Environmental Policy Centers at Harvard University, an associate member of Nuffield College, Oxford, and a member of the Advisory Panel of the President's National Commission for the Review of Antitrust Laws and Procedures. In addition to his position at the American Enterprise Institute as director of regulatory policy studies, Dr. Stelzer is an honorary fellow of the Centre for Socio-Legal Studies, Oxford, and political and economic columnist for the *Sunday Times* (London) and the *New York Post.* He is the coauthor of *The Antitrust Laws: A Primer* (AEI Press, 1998), now in its third edition.

Board of Trustees

Wilson H. Taylor, *Chairman*
Chairman and CEO
CIGNA Corporation

Tully M. Friedman, *Treasurer*
Tully M. Friedman & Fleicher, LLC

Joseph A. Cannon
Chairman and CEO
Geneva Steel Company

Dick Cheney
Chairman and CEO
Halliburton Company

Harlan Crow
Managing Partner
Crow Family Holdings

Christopher C. DeMuth
President
American Enterprise Institute

Steve Forbes
President and CEO
Forbes Inc.

Christopher B. Galvin
CEO
Motorola, Inc.

Harvey Golub
Chairman and CEO
American Express Company

Robert F. Greenhill
Chairman
Greenhill & Co., LLC

Roger Hertog
President and COO
Sanford C. Bernstein and Company

M. Douglas Ivester
Chairman and CEO
The Coca-Cola Company

Martin M. Koffel
Chairman and CEO
URS Corporation

Bruce Kovner
Chairman
Caxton Corporation

Kenneth L. Lay
Chairman and CEO
Enron Corp.

Marilyn Ware Lewis
Chairman
American Water Works Co., Inc.

The American Enterprise Institute for Public Policy Research

Founded in 1943, AEI is a nonpartisan, nonprofit, research and educational organization based in Washington, D. C. The Institute sponsors research, conducts seminars and conferences, and publishes books and periodicals.

AEI's research is carried out under three major programs: Economic Policy Studies; Foreign Policy and Defense Studies; and Social and Political Studies. The resident scholars and fellows listed in these pages are part of a network that also includes ninety adjunct scholars at leading universities throughout the United States and in several foreign countries.

The views expressed in AEI publications are those of the authors and do not necessarily reflect the views of the staff, advisory panels, officers, or trustees.

Craig O. McCaw
Chairman and CEO
Eagle River, Inc.

Paul H. O'Neill
Chairman and CEO
Aluminum Company of America

John E. Pepper
Chairman and CEO
The Procter & Gamble Company

George R. Roberts
Kohlberg Kravis Roberts & Co.

John W. Rowe
Chairman, President, and CEO
Unicom Corporation

Edward B. Rust, Jr.
President and CEO
State Farm Insurance Companies

John W. Snow
Chairman, President, and CEO
CSX Corporation

William S. Stavropoulos
Chairman and CEO
The Dow Chemical Company

Officers

Christopher C. DeMuth
President

David B. Gerson
Executive Vice President

John R. Bolton
Senior Vice President

Council of Academic Advisers

James Q. Wilson, *Chairman*
James A. Collins Professor of Management Emeritus
University of California at Los Angeles

Gertrude Himmelfarb
Distinguished Professor of History Emeritus
City University of New York

Samuel P. Huntington
Eaton Professor of the Science of Government
Harvard University

D. Gale Johnson
Eliakim Hastings Moore Distinguished Service Professor of Economics Emeritus
University of Chicago

William M. Landes
Clifton R. Musser Professor of Economics
University of Chicago Law School

Sam Peltzman
Sears Roebuck Professor of Economics and Financial Services
University of Chicago Graduate School of Business

Nelson W. Polsby
Professor of Political Science
University of California at Berkeley

George L. Priest
John M. Olin Professor of Law and
  Economics
Yale Law School

Thomas Sowell
Senior Fellow
Hoover Institution
Stanford University

Murray L. Weidenbaum
Mallinckrodt Distinguished
  University Professor
Washington University

Richard J. Zeckhauser
Frank Ramsey Professor of Political
  Economy
Kennedy School of Government
Harvard University

Research Staff

Leon Aron
Resident Scholar

Claude E. Barfield
Resident Scholar; Director, Science
  and Technology Policy Studies

Cynthia A. Beltz
Research Fellow

Walter Berns
Resident Scholar

Douglas J. Besharov
Resident Scholar

Robert H. Bork
John M. Olin Scholar in Legal Studies

Karlyn Bowman
Resident Fellow

John E. Calfee
Resident Scholar

Lynne V. Cheney
Senior Fellow

Dinesh D'Souza
John M. Olin Research Fellow

Nicholas N. Eberstadt
Visiting Scholar

Mark Falcoff
Resident Scholar

Gerald R. Ford
Distinguished Fellow

Murray F. Foss
Visiting Scholar

Diana Furchtgott-Roth
Assistant to the President and
  Resident Fellow

Suzanne Garment
Resident Scholar

Jeffrey Gedmin
Research Fellow

James K. Glassman
DeWitt Wallace–Reader's Digest
  Fellow

Robert A. Goldwin
Resident Scholar

Mark Groombridge
Abramson Fellow; Associate Director,
  Asian Studies

Robert W. Hahn
Resident Scholar

Kevin Hassett
Resident Scholar

Robert B. Helms
Resident Scholar; Director, Health
  Policy Studies

R. Glenn Hubbard
Visiting Scholar

James D. Johnston
Resident Fellow

Jeane J. Kirkpatrick
Senior Fellow; Director, Foreign
  and Defense Policy Studies

Marvin H. Kosters
Resident Scholar; Director,
  Economic Policy Studies

Irving Kristol
John M. Olin Distinguished Fellow

Dana Lane
Director of Publications

Michael A. Ledeen
Freedom Scholar

James Lilley
Resident Fellow

Clarisa Long
Abramson Fellow

Lawrence Lindsey
Arthur F. Burns Scholar in Economics

John H. Makin
Resident Scholar; Director, Fiscal
  Policy Studies

Allan H. Meltzer
Visiting Scholar

Joshua Muravchik
Resident Scholar

Charles Murray
Bradley Fellow

Michael Novak
George F. Jewett Scholar in Religion,
  Philosophy, and Public Policy;
  Director, Social and Political Studies

Norman J. Ornstein
Resident Scholar

Richard N. Perle
Resident Fellow

William Schneider
Resident Scholar

William Shew
Visiting Scholar

J. Gregory Sidak
F. K. Weyerhaeuser Fellow

Christina Hoff Sommers
W. H. Brady, Jr., Fellow

Herbert Stein
Senior Fellow

Irwin M. Stelzer
Resident Scholar; Director,
  Regulatory Policy Studies

Daniel Troy
Associate Scholar

Arthur Waldron
Director, Asian Studies

W. Allen Wallis
Resident Scholar

Ben J. Wattenberg
Senior Fellow

Carolyn L. Weaver
Resident Scholar; Director, Social
  Security and Pension Studies

Karl Zinsmeister
J. B. Fuqua Fellow; Editor, *The
  American Enterprise*

AEI STUDIES ON GLOBAL ENVIRONMENTAL POLICY
Irwin M. Stelzer, Series Editor

COSTS AND BENEFITS OF GREENHOUSE GAS REDUCTION
Thomas C. Schelling

THE ECONOMICS AND POLITICS OF CLIMATE CHANGE
Robert W. Hahn

MAKING ENVIRONMENTAL POLICY: TWO VIEWS
Irwin M. Stelzer and Paul R. Portney

www.ingramcontent.com/pod-product-compliance
Lightning Source LLC
Jackson TN
JSHW011944131224
75386JS00041B/1552